THE ENVIRONMENT

POLLUTION:

Causes, Effects

and

Mitigation Measures

VOLUME 1

ISAAC ARIGBEDE

(C) 2016 I.O. Arigbede

All rights reserved. No part of this publication may be reproduced, stored in a retrieval system or transmitted in any form or by any means without the prior permission of the author.

Printed by: AROLIS SERVICES,
              3, Sewage Avenue,
              Behind Jakande Estate,
              Abesan, Ipaja,
              Lagos State.

Phone: 07032550987, 07012987048.

PREFACE

Pollution is one of the major and most discussed global issues in the world today. Yet, billions of people did not understand this life-threatening global problem. Besides, all forms of vital discussions, wise suggestions and reasonable resolutions at world conferences achieve very little success among the participating nations. Why is it so?

THE ENVIRONMENT Pollution: causes, effects and mitigation measures volume one is written to enable the readers (the educated, the students, the professionals and the non-professionals) understand pollution and to educate them in the most simplified manner with a view to taking necessary actions for a healthy living in a clean environment.

This handbook has ten chapters and a total of forty six pages. This book discusses pollution extensively to enable readers have an in-depth knowledge of pollution; the different causes, the effects and the necessary measures required of individual, group, organizations and government with a view to preventing, reducing or controlling the severity of this global problem.

ACKNOWLEDGEMENT

This book has been made possible by the special grace of God Almighty who directed and guided me. I greatly appreciate the priceless gift of good health, precious time for this intensive research work, knowledge and wisdom God gave me in completing this work. I remain ever grateful to Almighty God.

I will also like to express my sincere appreciation to family and friends who contributed greatly to the successful completion of this book.

CONTENTS

	Pages
Preface	iii
Acknowledgement	iv
1. Introduction to Pollution	1 – 3
2. Types of Pollution	4 – 8
3. Human Causes of Air Pollution	9 – 12
4. Urban Air Pollution	13 – 16
5. Other Types of Urban Air Pollution	17 – 22
6. Mist, Fog, Haze and Smog	23 – 26
7. Acid Rain	27 – 30
8. Types of Water Pollution	31 – 36
9. Soil Pollution	37 – 40
10. Other Types of Pollution	41 – 46

CHAPTER 1

INTRODUCTION TO POLLUTION

The existence and survival of humans, animals and plants or vegetation depend on some highly precious and priceless natural provisions. Some of these natural gifts are air, water and soil together with all the substances that exist with or form the components of these natural gifts. To benefit fully from these natural gifts, humans, animals and plants are provided with all the necessary healthy body parts and systems, located in the proper habitat or environment and also provided with the ability to use the air, water and soil optimally and sustainably.

Unfortunately, the natural condition of air, water and soil or land have been disturbed greatly in many ways mainly by human activities and this unconcern attitudes continue to impact adversely on every living and non-living things globally. This fact is obvious in every part of the world and manifests itself physically as **pollution**.

What is pollution? What are the sources of pollution? What are the different types of pollution? What contaminants or pollutants are responsible for this global problem? What are the causes of pollution? How does it affect life and property on earth? Can pollution be prevented, reduced, controlled, or completely eradicated so that humans and animals can live a healthy and longer life free of avoidable calamities? This section will provide appropriate answers to these questions with a view to educating, creating awareness and enlightening readers in a simplified manner.

What is Pollution?

Pollution is a direct or indirect introduction of harmful or toxic contaminants, substances or pollutants either in the form of gases, liquids or solids into the air, water and soil or land, causing adverse changes and making life difficult for humans, animals and plants.

There are two categories or sources of pollution. These are point source and non-point source pollution.

What is Point Source Pollution?

Point source pollution is any single identifiable source of pollution from which pollutants are discharged, such as a sewage pipe, factory, ship or livestock farm. Some of the discharges from point source pollution are harmless but others are toxic to people and animals. The toxicity depends on the components of the discharge, its concentration, time of discharge, the weather conditions and where it was discharged.

What is Non-Point Source Pollution?

Non-point source pollution come from many sources, therefore it is difficult to trace the origin of the pollutants. Some of the origins of non-point source pollution include mining operation, rural and urban areas and so on. The pollutants from these sources are quite harmful and complex. Therefore, the problems created by non-point source pollution are difficult and expensive to rectify.

Most non-point source pollution occurs as a result of runoffs such as rainwater and melted ice. These runoffs flow into the soil and absorb any pollutants it comes into contact with and eventually empties into stream, river or ocean. Examples of non-point source pollutants include excess nutrients such as nitrogen and phosphorus from farmland, suspended sediments, pesticides and toxic chemicals, bacteria, viruses and trash.

CHAPTER 2

TYPES OF POLLUTION

There are different types of pollution in the world. Some of the types of pollution are air pollution, water pollution, soil pollution, litter pollution, noise pollution, visual pollution, light pollution, radioactive pollution, thermal pollution and plastic pollution. Among the listed examples, the greatest emphases are paid on three in most countries. These are air, water and soil pollution. This is due to their rapid spread within and outside its origin, ease of disease transmission across the nation and national boundaries, huge cost of rectifying problems, high cost of damage to humans, animals and property, high revenue loss to nations, complexity of problem, its disastrous effects and so on.

AIR POLLUTION

Air pollution is the presence or introduction into the air of substance(s) such as gases, dust, fumes (smoke) or, odor, which have harmful, poisonous and/or damaging effects on humans, animals, plants and property.

Any substances that contaminate the air and have adverse effects on living and non-living things are known as air pollutants. Examples of air pollutants are carbon monoxide, nitrogen dioxide, particulate matters, sulphur dioxide, Lead and other suspended organic or inorganic chemicals and non-chemical matters.

Causes of Air Pollution: There are two main causes of air pollution. These causes are natural and human.

Natural Causes of Air Pollution:

Natural causes of air pollution do not occur frequently and they cannot be prevented or controlled by humans. They can only be monitored and/or forecast. Sometimes sudden weather changes render forecast record useless. Some examples of natural causes of air pollution include but not limited to wind erosion, volcanic eruptions, evaporation of organic compounds and wildfires.

A. Wind Erosion:

Wind erosion is common in areas with wide open field such as desert, loose sandy soils, in regions with sparse vegetation, etc. During wind erosion, fine sand particles, dust, light materials and disease-causing germs from decayed matters are carried or transported from one area to another. This process is either mild and spread gradually or sudden and spread rapidly. The intensity of sun, the area covered, time, soil texture (smoothness or coarseness), soil structure and moisture content in the soil are some of the major determinants of the type and quantity of materials eroded. Most of the materials eroded remain suspended in the air for a long period of time because of their extremely low densities and they are transported over a long distance. This is an example and process of air pollution.

Effects of Wind Erosion:

Wind erosion is one of the carriers of natural air

pollutants and it has many adverse effects. The first among the problems of wind erosion is air pollution which affects living and non-living things within and outside its origin. Wind erosion blow away dry fertile soil surface and render the soil sterile. Wind erosion also cause serious damages to properties and huge economic losses to nations globally.

Besides, the greatest concern of air pollution caused by wind erosion is the terrible health effects on humans. Some of the health hazards of air pollution as a result of wind erosion include asthma, eye problem, food and water contamination and other strange bacteria, fungi and virus diseases.

How to Reduce Wind Erosion and its Health Hazards:

Wind erosion can be reduced by practicing good land management. Some of the practical methods include tree planting or forest development program, planting cover crops such as leguminous crops and creeping plants, regular irrigation of dry land, avoiding overgrazing of farmland. The health hazards can be reduced by using eye glasses, nose covers and staying away from wind-erosion-prone areas.

B. Evaporation of Organic Compounds:

Organic compounds are capable of contributing greatly to air pollution globally due to their characteristics. They have high vapor pressure at ordinary room temperature and low boiling point which make it possible for them to evaporate from the liquid or solid state to gaseous forms and enter the surrounding air or the atmosphere. Organic compounds are produced naturally and by human activities.

Causes of Organic Compounds:

Natural organic compounds are produced by plants (e.g. poplar, eucalyptus and oak trees), animals (e.g. ruminants), microbes (e.g. molds), volcanoes and fossil fuel deposits. The rate of evaporation of organic compound depends on sunlight for their formation and temperature for their evaporation and growth. Most of these organic compounds react with nitrogen oxides to produce ozone gas which manifest as smog in the atmosphere. Poplar trees and molds can release organic compounds into the atmosphere in large quantities.

Effects of Organic Compounds in Air:

Organic compounds can undergo different chemical reactions. This makes it very harmful on any objects they come into contact with. They can destroy vegetation. They can damage properties. They can kill animals when they eat plants containing chemical deposits. The chemicals can pollute rivers, seas and oceans and poisoned all living organisms in the water. Besides, they constitute great health

risks to humans. Some of the health hazards include skin diseases and itching, eye problems, cancer, damage to kidney, lungs, liver, fatigue, dizziness and so on. The type of disease caused depends on the type of chemical compound and the part of the body affected.

Necessary Measures:

The first and wisest step to take is to consult medical doctor when strange diseases are noticed. Contact the environmental protection agency regularly for air quality update. Stay away from intense sunlight.

C. Volcanoes and Wildfires:

Volcanoes and wildfires are other major natural causes of air pollution. These two factors will release so much carbon-rich compounds to the air. These compounds will remain in the air for a long period of time and cover a long distance. Among the compounds released are carbon monoxides (CO), carbon dioxide (CO_2), methane (CH_4), ash and other organic compounds.

Like the other natural causes of air pollution, the chemical compound released during volcanic eruption or wildfire will cause serious health hazards and damages to living and non-living things around and outside its area of origin.

CHAPTER 3

HUMAN CAUSES OF AIR POLLUTION

Pollution is not a new problem. Pollution of the environment or the atmosphere is as old as the world we all live in, the earth. The three most dangerous and most emphasized pollution namely air, water and soil pollution are not reducing ever since the creation of this world. Instead, it continues to aggravate, causing all forms of problems and disasters to life and losses to properties globally. Yes, there is natural pollution which is beyond human prevention or control. But the various activities of humans continue to add to the pollution problems and by extension, the greenhouse effect which ultimately lead to global warming/global climate change.

Among the three most common and most disastrous pollution, air pollution is topmost. This is a fact and it is still maintaining this scary leading position globally. The reasons are not far-fetched. This is due to human activities which include exploration, production, consumption, development and rapid population growth to mention just a few. What is air pollution? What are some of the human causes of air pollution?

What is Air Pollution?

Air pollution is the contamination of air with different types of substances that make it impure, hazardous and damaging to humans, animals, plants and properties.

Causes of Air Pollution:

Pollution can be indoor or outside. The area or location polluted determines whether pollution is indoor or outside.

Indoor Air Pollution

This is the primary source of air pollution. Little or no concern is shown in most homes and by most families. Babies and small children are exposed to serious and greater health hazards here. Indoor air pollution can occur in living apartments, offices, stores, warehouses, halls, cinemas, classrooms, gym, laboratories and any other confined, small and poorly ventilated spaces .

Causes of Indoor Air Pollution:

Some of the causes of indoor air pollution are; wall paints, furniture polish, synthetic materials, insecticides. Other causes include fumes from stoves, generators, heating coals, perfumes, cigarette or tobacco smoke, dead rats, molds from kitchen wastes and decayed foods, disinfectants, etc.

Effects of Indoor Air Pollution:

Indoor air pollution creates many health effects. Most of the adverse health effects are directly or indirectly related to respiratory diseases and breathing problems. Other health effects include food and drink poisoning, eyes and skin problems to mention just a few.

Prevention/ Control Measures:

Parents should create more space for increased ventilation. Reserve a room for keeping used, unused or surplus items. Seek advice from home hygiene and other relevant health professionals. Install window and door mosquito nets so as to keep all windows and doors open. Stay away from poorly ventilated or polluted areas. Buy and use less of synthetic materials. Check product compliance for environmental pollution standard before purchase. Always keep babies and small children out of the polluted area. Wash and cook food items properly. Follow product instructions and warnings. Consult your doctor if strange disease or health problems are observed.

Environmental/Outdoor Air Pollution

There are two main categories of environmental or outdoor air pollution. These categories are community or rural air pollution and urban, town or city air pollution.

Rural or Community Air Pollution

Rural areas are known to have low level of air pollution due to small number of or lack of industries and establishments which require high energy for operation. Most rural areas are either farming communities only, residential only or farming with residential houses. These types of rural settlements with low level of air pollution are common in many developed countries. However, the major air pollution problem here is the burning of large amount of fossil fuels such as petrol and diesel when moving from the community to the town or cities.

The situation is not the same in the developing countries. In other words, air pollution level is higher in these countries. Large amount of air pollutants are produced and released from many activities such as; extensive bush burning, use of firewood for all their cooking, use of toxic chemical items such as rubber, plastic, nylon and tires to burn rubbish.

Effects of Rural Air Pollution:

Rural dwellers experience similar health problems. Their case is even worse because of high level of poverty and illiteracy. Also, distance from the city or town and lack of infrastructure made it difficult for them to get access to clinics or hospitals. Hence, their lives are cut short most times by minor and curable health problems.

Necessary Measures:

Government should build hospitals with modern equipment and qualified medical personnel for those in rural areas. Environmental extension office and workers should be provided in rural areas to educate the people. Necessary infrastructure and renewable clean energy resource should be provided for them. This will help a lot in their farming operations.

CHAPTER 4

Urban Air Pollution

Urban area refers to towns and cities. Urban areas are characterized by high human population, small and big companies, many residential and industrial areas, offices availability of infrastructural facilities and continuous construction work, high level of transportation, etc. In view of the features of urban areas and the various activities which take place continually in it, it is obvious that air pollution will be an everyday occurrence leading to serious environmental problems. In fact, urban air pollution is the leading cause of man-made global warming.

Some of the human activities that directly or indirectly continue to produce and release air pollutants to the environment include extraction, refining or production and burning of fossil fuel products, manufacturing or production processes in industries, mining operations burning of household or garden wastes, agricultural activities, electricity or power plants operation and so on.

FOSSIL FUELS

What are fossil fuels?

Fossil fuels are natural energy reserves found in the soil, rock and sea or ocean depth. They are formed over millions of years from the decayed remains of animals and plants.

There are three types of fossil fuels. These are; oil, coal and natural gas. Fossil fuels are the greatest sources of air pollution globally.

It is a fact and it is logical to believe that the discovery of fossil fuels marked the beginning of pollution, and greenhouse effect which gives rise to what is known as global warming. The reason is because the entire human activities which involve the use of fossil fuels continue to produce and release toxic pollutants and greenhouse gases to the air. In other words, the extraction of fossil fuels, the refining and the burning or using of fossil fuels products by consumers all generate the greatest amount of toxic and non-toxic gases that is responsible for different types of pollution and global warming.

1. **Crude Oil:**

Some of the refined products of crude oil during the process of fractional distillation at varying temperature ranges include domestic and industrial **gases** such as methane, ethane, propane, and many others. **Petrol** or **gasoline** is the next product. This is used as fuel in small engine vehicles, firms, generators, small machines and devices. **Naphtha** is another product used in making variety of chemicals and chemical-based products.

Others are **Paraffin** or **Kerosene,** this is one of the commonest products of refined crude oil. It is used as fuel in jet engines, as heating oil, in stoves and lamps. **Diesel** is used as fuel in big vehicles, in power generators, in companies and for machine operations. **Fuel oil** is used as fuel for ships and very big engines. Wax and polish are some of the products derived from **lubricating oil.** Finally **Bitumen** residue is a major raw material for the road construction industries. It is used for surfacing roads.

The users of all these products generate a very large amount of one or combination of the following chemicals namely sulphur dioxide, carbon monoxide, carbon dioxide, nitrogen oxides, volatile organic compounds, ground ozone, particulates, lead and other heavy metals. All the gases and compounds listed here are major air pollutants and most of them are toxic.

2. Coal:

Coal is another category of fossil fuels. It is the most abundant among the three categories of fossil fuels. Coal is a compound fossil fuel. It consists of carbon, hydrogen, oxygen, nitrogen and sulphur chemical elements. These are the major components of raw coal ($C_{135}H_{95}O_9NS$) after mining. Coal production will yield varieties of simple and complex chemical products. Some of the products of coal processing are; **Ammonia,** it is used in the production of fertilizer. **Coal Gas** is used for industrial heating. **Coal Tar** is separated into various fractions and used as raw materials for the production of dyes, creosote, paint, pitch and so on. **Coke** is another vital product of coal processing. It is used in making iron and steel and also for industrial heating.

3. Natural Gas:

Methane is the main product from natural gas processing. Methane is a good clean energy. It is used for domestic, office and industrial work.

Like crude oil refined products, extraction or mining of coal and natural gas generate large amount of air pollutants in addition to greenhouse gases. Also like crude oil, these two fossil fuels are non-renewable and can cause serious health problems. Yet they are always in high demand due to its many uses by all industries, due to the variety of products that can be obtained from their processing, low price and availability in large quantities.

Necessary Measures:

Industries and consumers should fit catalytic converter to exhaust pipes and chimneys, they should use less fossil fuels products and more health-friendly clean energy; such as solar energy, wind energy and bio-fuels. Government should support the producers of renewable energy machines with low interest loans. This will enable them produce in large quantities for domestic and industrial uses at cheap prices. Government can combine with local and/or foreign investors to set up renewable energy firms. Government should include topics on pollution, global warming and renewable energy in school syllabus.

CHAPTER 5

OTHER TYPES OF URBAN AIR POLLUTION

1. BURNING AND DUMPING OF WASTES

Household and garden wastes in urban areas should be a source of serious concern for the public and the government because most of the wastes are not properly disposed of. In most homes, household and garden wastes are burnt regardless of the types and the chemical components of the wastes. Some of the household and garden wastes include paper, wooden materials, rubber, nylon, plastic, textile materials, dry grass, foam, rotten food and packaging foils.

This method of waste disposal always endangers life and property first in the closest neighborhood. If the waste is plenty and the burning continues, the pollutants could cover the areas for days. When household and garden wastes are burnt, toxic gases such as carbon monoxide, nitrous oxide, ashes, sulphur dioxide and other complex gases are released and these pollute the air.

Effects of Burning and Dumping of Wastes:

Burning of household and garden wastes constitute great health risk and affect the neighborhood and the entire environment. Different types of serious health problems namely respiratory diseases, eye burn and internal organ diseases are some of the adverse effects of burning household wastes inside or outside the compound. Refuse dumped in gutters and streets pollute air with offensive odor, spread disease germs, block drainage channels and cause flooding during rainfall.

Necessary Measures:

If burning of household wastes is allowed, it must conform to the rules and regulations of environmental protection agency. For example, household and garden wastes must be sorted into organic, compostable and recyclable matters. Sorted household wastes can be moved or sold to incinerator, compost and recycling plants owners for combustion or processing. Companies, home owners, tenants and every person must dispose of refuse properly. Companies should comply with environmental standards in packaging and materials used for their products. Consumers should adopt the concepts of reduce, re-use and re-cycle in their homes and buy environmental-compliant packaged products.

2. AGRICULTURAL ACTIVITIES

Agricultural practices or activities are compulsory for all nations as these provide food, fiber and other variety of raw materials for humans, animals and industries. The different types of agricultural practices play a crucial role in the type, quantity and effect of pollutant generated and released to the air. The main classes of agricultural production are livestock production, aquaculture, forestry and crop production. All types of specialization in agriculture are derived from these classes.

The management practices adopted in agriculture will have direct or indirect impact on the environment. If an agricultural establishment or farm uses more of chemical compounds, the impact on the environment will be much and vice versa. For example, high dependent on agro-chemicals such as bactericide, herbicide, nematicide, fungicide, inorganic fertilizer and other chemical inputs will result in the release of particulates and toxic gases into the atmosphere. If organic agriculture or organic farming is practiced, air pollution and other forms of pollution will reduce greatly.

Causes of Air Pollution:

The primary causes of air pollution in agriculture are the agro-chemical inputs used in both crop and animal production. Despite the usefulness of these chemicals, crop and animal production are affected adversely due to the concentration of pollutants in the air, particularly when aerial application is used. Besides, the farm workers are the primary victims of health problems caused by air pollution. This is followed by the people working or living in the vicinity of the farm and finally, the entire public.

Effects of Air Pollution by Agricultural Activities:

Agro-chemicals are complex and very harmful substances. Most of these chemicals are either in form of powder or particles. These chemicals can remain suspended in the air for a long period of time because of their low density. They can cause serious respiratory, eye, organ and skin diseases just to mention a few. It can also cause cancer if treatment is delayed. In fact, it could lead to death if an organ such as lung and kidney is affected and prompt medical attention is not provided.

Prevention/Control Measures:

Farms and agric industries should practice organic farming and depend less on chemicals. They should also practice biological control of pests and use improved seeds and seedlings for crop production. Livestock waste should be used as compost to produce biogas for energy and as a source of organic fertilizer for crops. Good soil management and soil tillage practices must be adopted to prevent wind erosion from polluting the air with dust.

3. ELECTRICITY OR POWER PLANT AIR POLLUTION

Electricity or power plant provides one of the most essential or compulsory services needed by any nation. Power plant operation is absolutely different from other establishments in many ways. Power plant operation involves the change or transformation of one form of energy to another form of energy. Power plant depends solely on the regular supply of large amount of fossil fuels namely petrol, coal and natural gas for transformation into electricity for domestic, industrial, office and many other uses.

Operational stages of all power plants consist of power generation, transmission and distribution. These three huge tasks require the use of fossil fuels both directly and indirectly depending on the stage of operation. During all these operational processes, abundant toxic pollutants are released to the atmosphere. Air pollution actually occurs when gases are released through long huge chimneys directly to the air.

Effects of Power Plant Air Pollution:

Most of the pollutants released during fossil fuel processing are also released by power plants during operation and service provision, because power plants are

fossil fuels dependent. Hence, power plant operation contributes greatly to all types of pollution and greenhouse effect. Without further listing, the health problems are the same with that of fossil fuel processing.

Prevention/ Control Measures:

Power plant engineers should use alternative renewable sources of fuel to generate, transmit and distribute electricity. Some renewable clean fuel options include hydro (dam construction), nuclear and solar energy. Power plants should recycle waste gases and stop gas flaring. Power plants should fit catalytic converter to chimneys and exhaust pipes to reduce emissions. Power plants should ensure regular maintenance of equipment. The public should be encouraged to install personal sources of energy generation such as solar and wind energy. Government should establish or assist companies to produce solar and wind energy components at cheap prices. Alternatively, government should support private investors with necessary facilities to establish solar and wind energy components firms.

CHAPTER 6

MIST, FOG, HAZE AND SMOG

MIST

Mist is a mass of fine droplets of water suspended in the atmosphere near or in contact with the earth limiting visibility. The density of water droplets is lower than fog and the visibility is higher than fog. It is possible for pilots to see more than 1,000 meters ahead and road users can see more than 200 meters ahead. Mist can be likened to the air (carbon dioxide) humans breathe out or exhale during winter which cools suddenly because of its contact with cold environment. Mist last for a short period of time.

FOG

Fog is a collection of water droplets or ice crystals suspended in the air or near the earth's surface. The density of water droplets is higher than mist and the visibility is lower than mist. Fog lasts for a longer period of time. Fog types include coastal fog, valley fog, radiation fog, upslope fog and so on. In foggy atmosphere, pilots cannot see more than 1,000 meters and motorist cannot see more than 200 meters.

HAZE

Haze refers to the atmospheric condition where dust, smoke and different dry particles obscure the clarity or visibility of the sky. In other words, haze occurs due to air pollutants. Most times, haze is noticed in areas far from where it originates because it is carried by wind to where it finally accumulates. Haze is brownish due to the high dust content.

SMOG

Smog is a type of air pollution which results from the combination of smoke and fog in the air. Smog is common in big industrialized cities, towns or countries with sunny, warm, dry climatic conditions. Some of the factors that initiate smog are high transportation, high or rapidly growing population, industrialization and so on.

Causes of Smog:

Smog is caused by all forms of human activities that lead to the release of air pollutants in a foggy atmosphere or environment. Expressed in another way, smog is caused when fog occurs in air-polluted environment or atmosphere. Some of the pollutants that are responsible for smog formation include primary pollutants such as nitrogen oxides, carbon monoxide, sulphur oxides, smokes and emissions from vehicles, industries and forest fires to mention just a few. Ozone (O_3) gas is a secondary pollutant caused by the reaction of hydrocarbon with oxides of nitrogen in the presence of sunlight. Smog can also be caused naturally by continuous strong volcanic eruptions. Smog is a complex harmful air pollutant. It is a non-point source pollutant.

Effects of Smog:

The numerous complex harmful pollutants which make up smog position this type of air pollution as a serious health problem. Yes, smog is common in highly industrialized nations but it can also spread across national boundaries causing different diseases. Some of the adverse effects of smog are; eye irritation, cough, asthma, heart diseases, lung diseases, cancer, and different respiratory diseases. Children and the elderly should be given priority attention because they are more vulnerable to diseases caused by smog.

Prevention/Control of Smog:

Always wear smog-gas mask. Stay away from smoggy areas. The public should be educated on the need to avoid polluting the environment. Laws should be made to punish individuals and the companies that disobey environmental rules. Establishments should contact the environmental protection agency regularly for update on air quality and operate their companies in compliance with global environmental standard. Officers of the environmental protection agency should also monitor the activities of the production companies and other establishments to ensure that they comply.

Government and/or private investors can make park and ride transportation arrangement to reduce emissions and air pollution. Neighbors and co-workers can arrange for car pool. Companies can also arrange for motor pool e.g. staff buses. This will also reduce emissions and pollution. These arrangements will greatly reduce smog formation that is caused by high vehicular movement.

CHAPTER 7

ACID RAIN

Mist, fog, haze and smog are the different media for acid rain. Mist and fog are natural weather conditions usually with reduced toxic and non-toxic matter or pollutants, depending on the location, natural and human activities. Haze and smog occur as a result of the combination of natural and human activities. These two have large amount of different toxic pollutants. The toxic effects of haze and smog become even more harmful and cause more destruction or damages to humans, land animals, soil organisms, aquatic ecosystem, plants or vegetation and properties when it rains.

Naturally, rain is pure and contains perfect or balanced chemical composition for healthy living of humans and animals. Also, pure rain contact with living and non-living things has non-destructive effects. Pure rain becomes acidic or acid rain when it combines with suspended toxic pollutants in the atmosphere. What is acid rain? What are the causes and the effects of acid rain? What control measures can be applied?

What is Acid Rain?

Precipitation (e. g, rainfall, snowfall, etc) with acid level or acid content that is below pH 5.5 is known as acid rain. Rain is acidic if it can have harmful and damaging effects on living and non-living things in the environment. PH is the scale used for measuring the amount or concentration of acid in a liquid or solution. The lower the pH level below 5.5 the higher the acid content in the rain the greater the damage it can cause.

Acid Rain Formation:

When the major acidic pollutants sulphur and nitrogen undergo series of separate chemical reactions starting with oxygen (air) and ending with water vapor to form mist, fog or dew then acid mist, acid fog or acid dew will be formed, if the pH level is below 5.5. But when these series of reactions ended with rain water instead of water vapor and pour as rain, then acid rain will be experienced. The others really occur but acid rain is common. The major types of strong acids that are formed include sulphurous acid, sulphuric acid, nitrous acid and nitric acid. Carbonic acid is also formed but it is a weak acid.

Causes of Acid Rain:

The major causes of acid rains are sulphur and nitrogen pollutants. These pollutants exist either in their free state or in combination with other chemical elements or compounds, especially with oxygen. These pollutants are released naturally and by human activities. Some of the natural ways include but not limited to lightning, volcanic eruption, soil organisms and forest fire. Human ways of releasing these pollutants include industrial operations such as production and combustion, power generation, mining, use of fossil fuel dependent vehicles, airplanes, ships, trains, machines and other devices. The level of population and industrialization are some of the major factors that determine the amount of sulphur and nitrogen pollutants released to the air.

Effects of Acid Rain:

Acid rain is liquid or solution. Hence, it will adversely affect anything it comes in contact with including the water table or underground water and cause huge losses. Acid rain will cause many health problems for humans and animals. These include eye irritation, respiratory problems, skin itching, damage to organs and most part of the body systems. It can also lead to death. This depends on the amount or acid concentration in the rain.

High concentrated acid rain through direct pouring or run off into the sea, ocean, rivers and lakes will result in rapid and complete death of living organisms, this is even more noticeable in lakes and ponds. Acid rain will cause large scale damage to soil, forest vegetation and agricultural crops. This will ultimately result in soil acidity, deforestation and poor or loss of agricultural crops with direct bad effects on food supply and the economy.

Acid rain will damage metal objects and properties that are in direct or indirect contact with it. In this case, corrosive and wearing effects can be seen on these metal objects and properties. Some of these include wearing of cement or concrete buildings, washing of building and car paints, corrosion of window frames, roofing sheets and water pipes. Also, transportation facilities such as trains, ships and buses will be affected. Acid rain will damage infrastructures such as bridges, roads, electric poles and masts. Acid rain can bleach textile materials.

Preventive/Control Measures:

The public should be educated on the causes, adverse effects and preventive measures for acid rain. Ponds, lakes and soil can be limed. Government should provide clean pipe-borne water for the public. Beverage companies must have modern quality control unit. Punitive measures should be applied for companies that fail to comply with the required health-friendly standard of operation, such as the use of chimney scrubbers, emission reduction, regular maintenance and repair of equipment. Government laboratories should perform regular water quality and acidity test before supplying the public. Metal objects and properties should be protected from rain. Consult medical doctors regularly for check- up.

CHAPTER 8

TYPES OF WATER POLLUTION

Water covers more than two third of the earth's surface. Life without water is absolutely impossible. Water is one of the compulsory or most needed life-sustaining requirements for all living creatures on land and inside water, for micro organisms and for plants. The numerous uses and the importance of water cannot be over-emphasized. Water is as old as the world itself.

Water is one of the most vital natural provisions with a very perfect chemical composition. It is provided to be used by humans, animals and plants for healthy living and growth. Water exists naturally in three different areas. These three areas are; **surface water**, namely oceans, sea, lakes and rivers. **Rock water** is found in rocks. And **underground water**, (water table). It is found at varying depths below the ground. Regular interactions between these three and the other natural phenomena makes possible a perfect water balance and water cycle.

But unfortunately, human activities continue to affect water adversely in form of pollution. Put in appropriate term, water pollution. What is water pollution? What are the causes of water pollution? How does polluted water adversely affect the environment? What measures can be taken to alleviate these negative effects?

What is Water Pollution?

Water pollution is the direct or indirect alteration of water purity/quality in their natural locations to the detriment of humans, animals, plants and properties. Water pollution can also be defined as the contamination of rivers, lakes, oceans, seas and underground water with substances that will affect the water quality and cause harm to the environment.

Water pollution commonly affects oceans, seas, rivers, and lakes as well as the underground water (water table) in many ways thereby posing danger to the environment. The level of water pollution can be determined mainly by the quantity of solid matter deposited, the concentration of toxic chemicals in the water and the ability of aquatic animals and plants to survive in the water.

Types of Water Pollution:

The liquid state of water makes it possible for polluted water to spread fast from one region or one location to another. This occurs both nationally and internationally or across national boundaries. There are different types of water pollution, and these depend among other things on the cause of pollution, where the pollution happened and the types of pollutants involved. Some of the common types of water pollution include nutrients pollution, surface water pollution, ground water pollution, chemical water pollution, thermal pollution, suspended matter pollution and so on.

1. Nutrients Pollution:

Nutrients pollution is associated with agricultural or home gardening activities. This occurs when excess and dissolved plant nutrients are washed by runoff into rivers, lakes and oceans or when excess nutrients are drained into the underground water. Some of the plant nutrients include nitrogen and phosphorus. The drainage of plant nutrients is aided by rainfall. The level of pollution here depends on the area of farmland or garden where plant nutrient was applied, quantity of unused nutrients by the plants, soil type and so on.

2. Surface Water Pollution:

Surface water pollution refers to the contamination of water bodies namely seas, lakes, rivers and oceans. Surface water pollution is the easiest and the fastest means of water pollution due to the wide open surface and the continual rapid flow of water. This type of pollution is non-point source pollution as the pollutants come from many different sources. Besides, surface water pollution allows all forms of pollutants such as liquid, viscous, small and large waste materials. Surface water pollution is common in industrialized and highly populated areas, towns, cities and coastal areas.

3. Ground Water Pollution:

Ground water pollution occurs when liquid pollutants leak through the soil to the water table and aquifers. Chemicals, sewage water, household waste water, acid rain and light oils or oils that are melted at high temperature are some of the pollutants which cause underground water pollution. The amount and type of rainfall, the speed and amount of surface runoff, concentration of pollutants, source of pollutants, soil type and the soil gradient are some of the factors which determine the rate of drainage of pollutants to the water table and the extent of ground water pollution.

4. Chemical Water Pollution:

Water pollution that involves chemicals only is known as chemical water pollution. This type of water pollution is the most harmful. This is because it is difficult to rectified, it can mix without being noticed and still be very harmful to humans, animals and plants both on land and in water. Some of these chemicals include organic and inorganic liquids from industries.

5. Thermal Pollution:

When the temperature of a surface water body is raised abnormally as a result of human activity causing adverse effects on aquatic life, the process is called thermal pollution. For example, when industries or companies use water as coolant and pumped the hot water back, the temperature of the surface water is increased abnormally. Runoff from urban areas can also cause thermal pollution if it is deposited in surface water. Sudden increase and sudden decrease in water temperature known as thermal shock is a good example of thermal pollution.

Causes of Water Pollution:

There are many causes of water pollution. Some of these causes are; sewage and waste water, septic tank, runoff, ocean and marine dumping, underground oil and chemical storage tanks and pipe leakages, acid rain, air pollutants, industrial wastes dumping, radioactive waste, heavy metals, harmful algae, oil tankers and so on.

Effects of water pollution:

Water pollution has many adverse effects on the environment. The adverse effect of water pollution is more obvious in aquatic animals. Death of aquatic plants and animals, disruption of food chains and ecosystem, decrease in the level of dissolved oxygen by harmful algae, water-borne diseases, poisoning and death of humans and animals that use or drink from the water and economic loss are some of the bad effects of water pollution.

Prevention:

Industries and manufacturing companies must build cooling ponds and cooling towers for their machines instead of using sea, river or lake. Government should ensure that all companies comply with environmental laws. Environmental sanitation day(s) can be observed in states and in the country as a whole. Charges should be imposed on industries that fail to comply with environmental requirements. Waste heat can be recycled for other uses. Regular water quality tests must be performed and the results must be made known to the public. Some examples of water quality tests or analyses that can be performed include biological, chemical, physical and radiological water quality analyses.

CHAPTER 9

SOIL POLLUTION

Soil formation results from the gradual process of breakdown of rocks. These processes include biological, chemical and physical weathering, erosion and volcanic eruptions. Soil consists of different components. These are organic matter, minerals, water and air. These components are found in varying amounts in soil depending on the process of soil formation. Hence, there are different types of soil namely clay, silt and sand soils.

Soil is the natural supplier of nutrients for plants which provide food for humans and animals. The type of soil in an area determines the types of food crops that will be cultivated in the area. Good quality fertile soil will result in different crops and high yield. The survival, population and continual existence of all living things and plants in a place or a region depend greatly on the soil fertility. For example, desert is without people and plants.

Soil cover less than one third of the earth's surface. The cultivable portion of soil is currently providing over seven billion people with food, and is still capable of serving even much greater population of humans with the needed food (if good soil management is practiced). In addition to this very large population of domestic and wild animals in the world also depend on the crops from the soil.

Apart from the primary importance of soil for the production of food crops and horticulture, soil is very vital for forest growth and development. This serves as home to wild animals and fruits, trees for various uses and revenue generation for government.

From the aforementioned, it is wise to believe that the size of fertile land in the world is capable of feeding the ever growing population of mankind and animals if it is properly managed. But it is very sad to say that the very little portion of the fertile land that is cultivated and the virgin soil are badly managed by human activities in so many ways. Because of the destructive human activities on the soil, some regions are already having very serious food shortage problem. If care is not taken, this may escalate into greater global food problem. The totality of human activities that continue to pose serious problems for mankind, animals and plants or the entire world will also adversely affect the soil. What is soil pollution? What are some of the causes of soil pollution? How does soil pollution affect the environment? What are the measures that can be taken to deal with this problem?

What is soil pollution?

Soil pollution is the alteration of soil quality and soil ecosystem (mainly by chemicals or liquids) due to human activities to the extent that soil use becomes harmful directly or indirectly to humans, animals and plants. Soil pollution is also the presence of toxic chemicals in soil in high concentrations that will be harmful to living things, plants and the soil ecosystem.

Causes of Soil Pollution:

Soil pollution occurs when toxic liquids or toxic chemicals drain into the soil. Solid wastes pollute the soil surface or land initially. After complete decomposition, toxic liquid from the decomposed matter is finally washed into the soil by rain. Unlike liquids and chemicals, soil pollution by solid wastes takes time and is gradual. Soil pollution is caused by many pollutants from different sources. Some of the common pollutants include but not limited to petroleum products refining, agriculture chemicals, nuclear wastes, toxic particles carried by wind and water erosion, buried wastes, heavy chemicals, industrial waste, wastes water from construction sites, acid rain, sewage effluents, oil spills and urban runoff.

Effects of Soil Pollution:

The fact that soil is the engine room for food for all living things and plants means that its pollution will result in great disaster for humans and animals. This is because crops and other plants will reduce or cease to grow and result in food shortage for humans and animals. This will eventually result in loss of many lives. Also soil pollution will reduce soil fertility as toxic pollutants will react with soil nutrients and form complex compounds that will not be useful for plant growth. Waste water from construction sites will block pore spaces in soil and also cement or bind soil particles thereby destroying the soil structure. Large amount of soil pollutants can be washed down by heavy rainfall and pollute underground water. Waste water or runoff from industries and urban areas will pollute soil if there are no good drainage channels.

Excessive use of agricultural chemicals and inorganic fertilizers will adversely affect microbial activities, affect soil nutrient balance and even affect humans and animals that feed on the crops harvested from the soil.

Prevention/Control Measures:

Government agents should educate the public on proper wastes (particularly liquids and chemicals) disposal and also ensure industries comply with rules and regulations. Chemical, oil and gas pipes should be protected from been vandalized to avoid leakage into the soil. Detail soil quality test should be performed before cultivating land for crop production. Farmers should use less inorganic fertilizer or practice organic farming. Shifting cultivation or land rotation can be practiced where fertile land is surplus. Harvested farm produce and meat (pork, beef, mutton, turkey, chicken, etc.) should be tested for chemical contamination and certified fit for consumption before transfer to the stores and markets.

CHAPTER 10

OTHER TYPES OF POLLUTION

1. NOISE POLLUTION

Noise pollution is another type of pollution. It refers to any sound that is either abnormal, harmful, causing discomfort, annoying or generally disturbs people and animals in the environment.

Causes of noise pollution:

Some of the causes of noise pollution include sporting events in sport centers, musical performance from entertainment venues, loud speakers from record stores, uncontrolled gatherings, motor vehicles, airplanes, rock blasting in mining sites, factory operation, etc.

Effects of noise pollution:

Like other types of pollution, noise pollution has very serious health problems. Human ear has noise limit. When this limit is exceeded due to noise pollution, it can cause temporary or permanent damage to the ear drum which may eventually lead to deafness. Babies and small children are more vulnerable to these adverse effects. Noise pollution can cause stress, it can cause shock particularly for the elderly and small children. It can cause lack of concentration. It can affect buildings. Noise pollution can also cause deafness in animals.

Prevention/Control of noise pollution:

Enforcement of rules and regulations and punitive measures is one of the best controls of noise pollution. Noise-free day(s) can also be observed. Warning sign boards should be used. It is wise to walk away from noise polluted sources to avoid hearing impairment. Noise-polluting companies such as aviation should be located far from residential and densely populated areas.

2. RADIOACTIVE POLLUTION

Radioactive pollution is the contamination of the environment by harmful rays from nuclear plants and radioactive materials.

This type of pollution is common in developed countries where nuclear plants are used and where nuclear materials are mined and manufactured. Radioactive pollution is very dangerous when it occurs.

Causes of radioactive pollution:

Radioactive pollution is caused by the release of radioactive pollutants by bomb explosion, nuclear plants, manufacturing companies, malfunctioning of nuclear plants, war, mining of radioactive raw materials, improper disposal of nuclear waste e.g. Uranyl nitrate.

Effects of radioactive pollution:

Many of the adverse health problems of radioactive pollution do not manifest immediately and these may not be detected if there is no regular medical check. Some of the health problems of radioactive pollution are; skin problems, cancer, defect on fetus and babies. Also, it can cause damage to crops and vegetation. It can pollute soil and water. Like humans, many animals also suffer from the adverse health effects of radioactive pollution.

Prevention/Control of radioactive pollution:

Nuclear power plants, mining and manufacturing companies must be located far away from towns, cities and residential areas. Producers and users must adopt good containment methods. Separate infrastructure and facility must be provided for these companies to avoid contaminating public utilities. Strict rules and regulations must be made and enforced by government agencies.

3. LIGHT POLLUTION

Light pollution is the excess or abnormal man-made lighting such that it becomes a nuisance, harmful to humans, animals and the environment.

Light is very important for humans, animals and for the environment as a whole. It is very useful for humans and animals as it brightens and helps the sight in darkness in the evening or during the night. Availability of light makes it possible for humans and animals to perform many activities when darkness falls. Light pollution can occur from one source such as high voltage electric bulb or from many sources such as an electric bulb and motor vehicle light in streets or in open areas.

Causes of light pollution:

Mix blinking colored light in disco hall is one good example of light pollution. Excessively bright lighting in cities, public gatherings and street is also light pollution. Motor vehicles cause light pollution when full light is used without any need for it. Advertisement board can cause light pollution if very high voltage is used.

Effects of light pollution:

Excess lighting can cause temporary or permanent blindness for humans and animals. Babies and small children are more vulnerable. It can damage the retina. It can also strain the eyes. Light pollution can cause headache. Excessively bright light from high wattage bulbs can cause increase in body temperature. It can damage some properties.

Prevention/Control measures:

Energy saving power sources is a good prevention or control measure. This will generate less heat and has no damaging effects on the eyes. The public should be educated on the need to avoid light pollution. Offenders may be charged for causing light pollution. Use of Dark eyeglasses is a wise method of avoiding light pollution. It also makes sense to walk away from the source of light pollutant. Do not keep babies and small children near excessively bright or high voltage light source.

4. PLASTIC POLLUTION

Plastic is a vital material used at homes and by many industries for variety of products. Also many companies produce different types of plastic products. Plastic is competing with paper in many industries nowadays. This is because it can be recycled and reused in many forms, it can be molded, it is corrosion-resistant, the raw material is abundant and it is cheaper and lighter to produce than metallic materials. Despite all the uses and importance of plastic and plastic materials, there is a growing trend in the handling of plastic in many countries and this continue to cause what is known as plastic pollution. Plastic pollution is an example of land or litter pollution.

What is plastic pollution?

It is the improper disposal of used or surplus plastics in offices, homes and industries such that it becomes a nuisance particularly on land and water surfaces.

Causes of plastic pollution:

Plastic pollution emanates from human activities. The greater part of plastic pollution is caused by the users. While most companies will recycle plastic or plastic materials in their premises most consumers will throw away or dispose plastic wastes carelessly on the land (in gutters, roads, streets, landfills, event centers etc.) and in water bodies such as rivers, oceans and seas. Plastic pollutants are used plastics in different sizes, shapes color and weights with or without chemical odor.

Effects of plastic pollution:

Improper disposal of plastic waste in large quantities will block drainage, affect water navigation, disrupts business and economic activities in coastal areas and harbors, can be swallowed by some aquatic animals and cause digestive or intestinal problems. It can also choke and lead to the death of some aquatic animals. Poor quality plastic or plastic materials with high chemical odor can cause diseases in humans. Plastics are flammable, it can cause fire outbreak if placed near naked fire. Plastics are not biodegradable (cannot be destroyed by microbes). Broken plastics can serve as breeding sites for insects. It can affect soil structure and crop production.

Prevention/Control of plastic pollution:

The public should be educated on the proper disposal of plastic wastes. Plastic materials should be reused and or recycled. Companies should encourage users to return used plastic wastes in exchange for products or money. Companies should employ workers to collect used plastics and plastic materials for recycling. Environmental agents should ensure companies use more of biodegradable items or materials for their product packaging. Laws should be made to make companies comply with international requirements on the use and production of plastic and plastic materials.

www.ingramcontent.com/pod-product-compliance
Lightning Source LLC
Chambersburg PA
CBHW070412190526
45169CB00003B/1226